Animals on the Farm

Horses

by Cari Meister

Bullfrog Books

Ideas for Parents and Teachers

Bullfrog Books give children practice reading informational texts at the earliest reading levels. Repetition, familiar words, and photo labels support early readers.

Before Reading
- Discuss the cover photo. What does it tell them?
- Look at the picture glossary together. Read and discuss the words.

Read the Book
- "Walk" through the book and look at the photos. Let the child ask questions. Point out the photo labels.
- Read the book to the child, or have him or her read independently.

After Reading
- Prompt the child to think more. Ask: Would you like to ride a horse? Would you like a horse as a pet? Why or why not?

Bullfrog Books are published by Jump!
5357 Penn Avenue South
Minneapolis, MN 55419
www.jumplibrary.com

Copyright © 2017 Jump! International copyright reserved in all countries. No part of this book may be reproduced in any form without written permission from the publisher.

Library of Congress Cataloging-in-Publication Data
Meister, Cari.
Horses / by Cari Meister.
    p. cm. —(Bullfrog books: animals on the farm)
Includes index.
Includes bibliographical references and index.
Summary: "A horse narrates this photo-illustrated book describing the body parts and behavior of horses on a farm. Includes picture glossary."
—Provided by publisher.
ISBN 978-1-62031-004-5 (hardcover : alk. paper)
ISBN 978-1-62031-631-3 (paperback)
1. Horses—Juvenile literature. 2. Horses—Behavior—Juvenile literature. I. Title.
SF302.M34 2013
636.1—dc23
                                            2012008421

Photo Credits: Alamy, 8, 17, 23tr; Dreamstime, 1, 3, 4, 10, 24; Getty, 16; iStockphoto, 3, 15; Shutterstock, 5, 8-9, 14, 19, 20t, 20b, 21, 22, 23br, 23bl; Superstock, cover, 6-7, 11, 12-13, 23tl

Series Editor: Rebecca Glaser
Series Designer: Ellen Huber
Production: Chelsey Luther

Printed in the United States of America at Corporate Graphics in North Mankato, Minnesota

Table of Contents

Horses on the Farm	4
Parts of a Horse	22
Picture Glossary	23
Index	24
To Learn More	24

Horses on the Farm

I am a horse.
I live on a farm.
Have you ever
seen a horse?

Do you see my friends?

Horses like to live in herds.

Do you see my long tail?

It swishes away bugs.

Do you see my ears? When they are up, I am happy.

When they are back, he is mad.

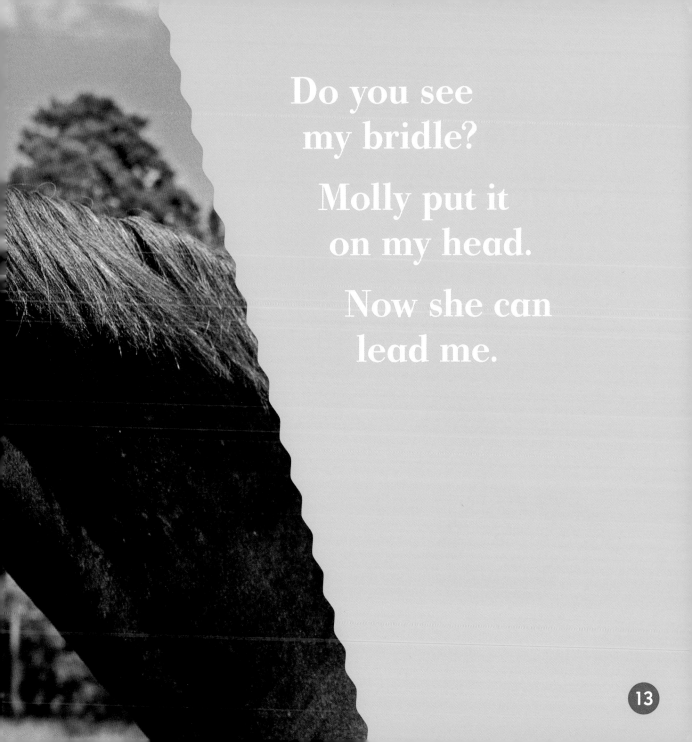

Do you see my bridle?

Molly put it on my head.

Now she can lead me.

Do you see my soft hair?

Ben uses a curry comb to clean it.

curry comb

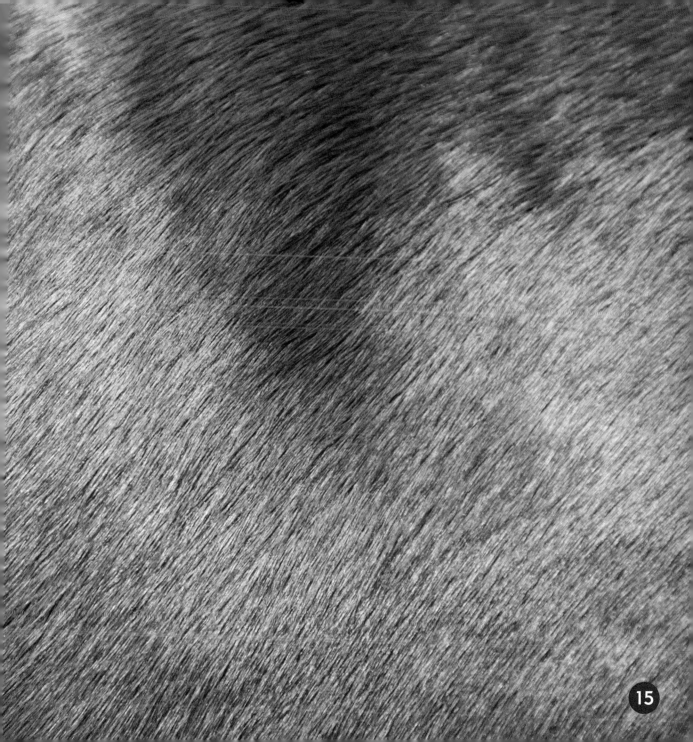

Do you see my hooves?

Molly uses a hoof pick to clean them.

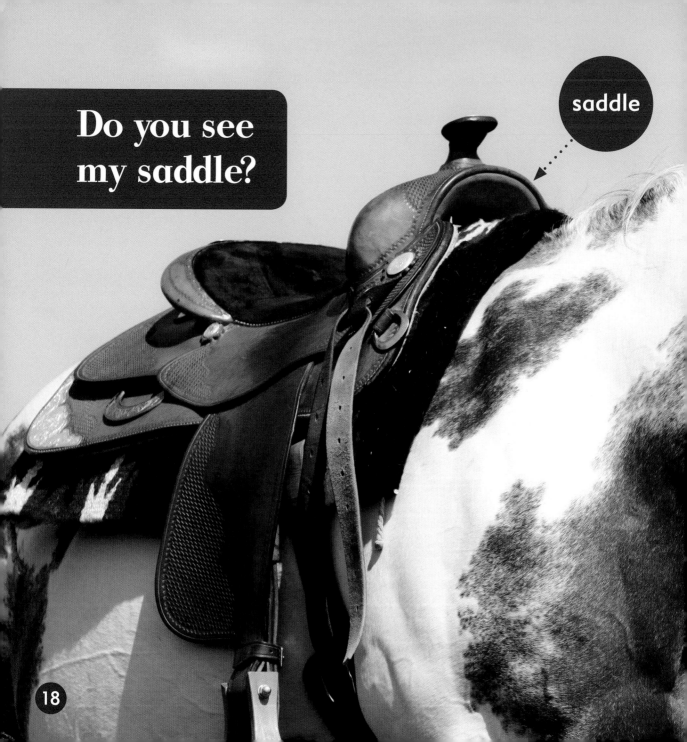

I am ready.

Let's go for a ride!

Parts of a Horse

ears
Horses can show feelings by how their ears point.

mane
Long, thick hair on the head and neck of a horse.

tail
Horses use their tails to swish away flies.

hoof
The hard outer covering of an animal's foot.

Picture Glossary

bridle
A head strap used to lead a horse.

hoof pick
A tool used to clean a horse's feet.

curry comb
A grooming tool with nubs, used in circle motions to clean a horse's hair.

saddle
A leather seat for a rider on the back of a horse.

Index

bridle 13	herds 7	saddle 18
curry comb 14	hoof pick 17	tail 8
ears 10, 11	hooves 16	
hair 14	ride 20	

To Learn More

Learning more is as easy as 1, 2, 3.

1) Go to www.factsurfer.com

2) Enter "horse" into the search box.

3) Click the "Surf" button to see a list of websites.

With factsurfer.com, finding more information is just a click away.